Rhayssa Rodrigues Muzi

Analysing the feasibility of using solar energy in an on-grid system

Rhayssa Rodrigues Muzi

Analysing the feasibility of using solar energy in an on-grid system

Supply alternative in a model home

ScienciaScripts

Imprint

Any brand names and product names mentioned in this book are subject to trademark, brand or patent protection and are trademarks or registered trademarks of their respective holders. The use of brand names, product names, common names, trade names, product descriptions etc. even without a particular marking in this work is in no way to be construed to mean that such names may be regarded as unrestricted in respect of trademark and brand protection legislation and could thus be used by anyone.

Cover image: www.ingimage.com

This book is a translation from the original published under ISBN 978-613-9-72028-6.

Publisher:
Sciencia Scripts
is a trademark of
Dodo Books Indian Ocean Ltd. and OmniScriptum S.R.L publishing group

120 High Road, East Finchley, London, N2 9ED, United Kingdom
Str. Armeneasca 28/1, office 1, Chisinau MD-2012, Republic of Moldova, Europe
Printed at: see last page
ISBN: 978-620-7-96127-6

DEDICATORY

I dedicate this work to my parents, siblings and aunt Iêda (in memoriam) who always dreamed of seeing me graduate and is unfortunately not here today to celebrate the realisation of this dream with me.

ACKNOWLEDGEMENTS

I would firstly like to thank God, who comes to my aid in times of anguish, without whom none of this would be possible.

To my parents and siblings who, with great affection and support, spared no effort to help me reach this stage of my life.

To my friends, for the joys and sorrows shared.

To all the teachers and coordinators of the course, who in one way or another have been important in my academic and personal life, not forgetting the UFMT teachers who accompanied me on my first steps of this long journey.

To all those who have contributed in some way to my education.

SUMMARY

The growing demand for electricity is one of humanity's greatest challenges and in the quest to overcome this, a new challenge has arisen: generating electricity with the lowest possible emissions of pollutants. As a result, investment in renewable energies has been growing in Brazil and around the world. Brazil has a privileged geographical location that provides natural conditions for harnessing renewable sources. In this scenario, solar energy stands out. Photovoltaic generation depends, among other factors, on solar radiation, a factor that cannot be controlled and is highly variable due to climatic conditions. This paper presents a case study of photovoltaic electricity generation in the town of Iúna - ES, showing that despite the variables imposed by nature itself and despite the fact that it is still an expensive energy to install, it is feasible to use solar energy to generate electricity, because even in the face of negative factors such as those mentioned above, it is possible to generate electricity even in periods of lower radiation and this generation reduces the energy consumption of the local utility, thus reducing energy costs, making the return on investment happen long before the useful life of the photovoltaic system is over.

Keywords: Solar energy; Photovoltaic generation; Feasibility.

INTRODUCTION

One of the main basic requirements of citizenship in today's world is access to electricity. Without it, human beings are marginalised from development.

Thus, one of humanity's greatest challenges in this century is to bring electricity to the 1.3 billion people around the world who still don't have access to it, totalling 18% of the world's population. Brazil is a little better off compared to the rest of the world, as it is estimated that only 6 per cent of the Brazilian population still lives without electricity (ACTION4ENERGY, 2015).

Figure 1: Map of where people without access to energy live.
Source: (ACTION4ENERGY, 2015)

Another important challenge when talking about electricity is to produce it with the lowest possible emissions of pollutants, in order to avoid aggravating environmental problems at local and global level. From this point of view, solar energy is one of the most promising sources of sustainable energy generation.

And in this quest for the harmonious and appropriate use of natural resources, the idea is to make the electricity chain more efficient, from generation to utilisation, passing through each of its stages (transmission and distribution).

distribution), seeking to ensure that this interaction is balanced with the environment in its broadest formulation, i.e. that the process is sustainable in its broadest concept (REIS, 2011).

Throughout the last century, the main supply of energy was obtained mainly from fossil fuels such as oil and coal, supporting economic growth and its respective transformations. However, all this growth has come from energy sources that are harmful to the environment, causing notable environmental damage, especially the contribution of carbon dioxide emissions and other greenhouse gases. However, at the beginning of the 21st century there was a need for sustainable development models for energy generation (BIGGI, 2013).

These sources of energy that have been widely used in the last century are non-renewable sources of energy, which means that one day they will run out. "World oil reserves total 1,147.80 billion barrels and annual consumption of this fossil fuel is estimated at 80 million barrels/day" (RATHMANN et al., 2005, p. 3). This leads to the conclusion that by 2046 the world's oil reserves will be exhausted. It is very important to emphasise that this calculation did not estimate the trend of increasing consumption, which leads us to conclude that if there are no new discoveries of oil reserves, it will run out even earlier than estimated. It is also worth remembering that, as oil is used up and no new reserves are discovered, the supply will get smaller and smaller, causing the price of oil to rise to such an extent that its use will no longer be interesting, the law of supply and demand, highlighting the need to find alternative sources of energy resources to replace oil (RATHMANN et al., 2005).

Figure 2 shows the proven oil reserves in billions of barrels by geographical region in the world in 2014, as can be seen.

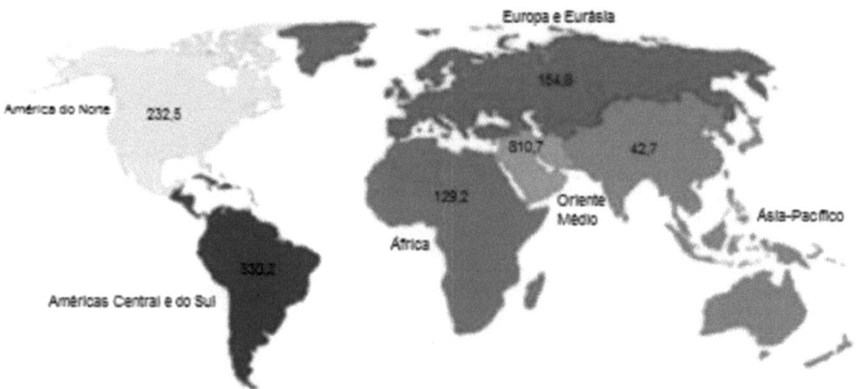

Figure 2: Proven oil reserves, by geographic region - 2014 (billion barrels)
Source: (Adapted from BP STATISTICAL REVIEW OF WORLD ENERGY, 2015)

When it comes to harnessing renewable energy sources, Brazil's geographical location is privileged and provides natural conditions for harnessing renewable sources in almost all of its regions due to its continental dimensions, with a vast stretch of coastline, intense solar resources, large hydrographic basins and an abundance of plants. In this race to find renewable energy alternatives to make up each country's energy matrix, Brazil is at an advantage, given that around 87% of the country's primary energy matrix is renewable. However, 65 per cent of this energy matrix comes from hydroelectric plants, leaving Brazil on the brink of collapse, as this type of source depends exclusively on rainfall and due to the low level of reservoirs in the Southeast, Brazil is in a delicate position (Adapted from BIGGI, 2013).

OBJECTIVES

GENERAL OBJECTIVE

- The aim of this work is to analyse the use of solar energy, bearing in mind that this energy is clean and renewable and has great potential to make up the energy matrix of most existing countries, making it a great alternative for countries that currently rely on non-renewable sources or even for countries like Brazil, which despite having a large percentage of its energy matrix coming from renewable sources, depends on factors such as rain, which is not constant, and the sun.

SPECIFIC OBJECTIVES

- Analysis of the use of solar energy within a model project for a typical Brazilian home (family of 5), highlighting the benefits of the feasibility study and the impacts of its adherence in Brazil.

CHAPTER 1

THEORETICAL BACKGROUND

1.1 RENEWABLE ENERGIES

For some decades now, the search for a sustainable development model and its subsequent implementation has been taking place, based on the organisation of society and prompted by the various environmental and social problems arising from the use of dirty energy, which is energy that causes damage to the environment or society, such as global warming and the poor distribution of natural and human wealth. This discussion has led to a growing awareness of the considerable interference that human systems impose on natural systems.

In this way, the use of the sustainable development model can contribute not only to overcoming current problems, but can also guarantee life itself by protecting and maintaining the natural systems that make it possible. For this to happen, profound changes are needed in the current systems of production, the organisation of human society and the use of natural resources.

The global discussion of the sustainable development model began in the 1970s and continues today, from an increasingly broad and participatory perspective. Since the first discussions, numerous discussion forums have been set up and various environmental agreements have been negotiated with the aim of reviewing the economic model adopted for development and curbing the trend towards the exhaustion of natural resources (REIS; FADIGAS; CARVALHO, 2005).

> Within this vision, a system based on the rational use of renewable resources, the recycling of materials, the fair distribution of natural resources and respect for all forms of life, offers a solution with a dynamic and harmonious balance between human beings and nature (REIS; FADIGAS; CARVALHO, 2005, p. 8).

Based on these general characteristics that are necessary for the new prototype, energy strategies and policies for sustainable development must be defined. Energy production, transport and use must therefore be re-evaluated, and energy planning must be rethought in order to add new technologies and methods, management practices, usage habits and the involvement of the population, without forgetting to take into account the reality and degree of development of each country before any choice is made. In view of this, the question that arises is how to find the appropriate paths within each specific context, and how to lay a solid foundation for continuing with the changes that will lead us to sustainable development (REIS; FADIGAS; CARVALHO, 2005).

There are several other major issues on the world stage that pose challenges to the practice of sustainable development. However, it is not the aim of this paper to go into them in depth. What we want to talk about here is the challenge of using one of the infinite natural resources that exists to generate energy with the practice of sustainable development, in this case solar energy.

1.2 SOLAR ENERGY

Solar energy, along with wind energy, is one of the most important new renewable technologies for Brazil, as it is more likely to be applied in the short term and can be used both for synchronous operation with a power system and to supply an isolated system.

When used in parallel with the grid, their application is closely linked to technological and economic aspects. In the case of isolated systems, these forms of energy generally compete with the extension of the electricity grid and are often more advantageous.

Over time, the cost of solar photovoltaic generation has been falling and shows signs of potential for further reduction, due to the scale factor when this type of generation becomes more widespread, given that the availability of the sun is practically universal. In the search for a more efficient integrated application of electricity, the use of specialised solar-photovoltaic panels in homes and buildings, together with automation systems and operating in parallel with the grid, has been the subject of several pilot projects (REIS, 2011).

The massive use of this form of energy generation in some more developed countries is already a reality. Germany, for example, already produces half of its energy needs through solar panels alone, as seen in figure 3, because the country lives with less sunlight than 90 per cent of the rest of the world's population and with a huge energy requirement since it is an industrial country (SOLAR PHOTOVOLTAIC PANELS, 2016 a).

Figure 3: Sunroof houses in Freibourg, Germany
Source: (SOLAR PHOTOVOLTAIC PANELS, 2016).

1.3 SOLAR ENERGY GENERATION

There are two ways of generating electricity using solar energy. They are:

1[a] Indirectly, in what are known as thermosolar systems, in which during the generation process they are indirectly converted into electrical energy through various technologies based on the use of solar energy to provide thermal energy which is then used to generate electricity.

2a Directly, through the use of photovoltaic panels.

There are several solar thermal systems installed around the world, some of them in the form of pilot projects and others in commercial operation. These systems are implemented in regions with a high incidence of solar energy and there are studies visualising their installation in large-capacity systems in desert areas, related to long-distance transmission. Despite the fact that unit energy costs are still high, there are currently a number of pilot research projects around the world and the fact that it is an energy source makes it an extremely important option because it is the most widely available renewable source in the world.

Solar photovoltaic generation, on the other hand, although smaller, has been much more widely used not only in developed countries, but also in developing countries, especially for powering small isolated systems, electrifying solitary equipment and in pilot projects. Although its cost is still not very attractive, it is rapidly decreasing as it becomes more widely used and as technology evolves. In the medium and long term, solar photovoltaic generation can be considered a very attractive form of energy generation not only for Brazil but also for the whole world (REIS, 2011).

1.4 COST OF SOLAR ENERGY

The price of solar energy, instead of being compared to the price offered by the generating plant, is calculated and compared to the price paid by final residential consumers. When compared to other energy generating sources, solar energy is considered to be very expensive and can be up to 50 times the cost of setting up a small hydroelectric power station, but the cost of the energy generated over the system's useful life, which is approximately 30 years, is only 10 times higher for isolated systems and for generation connected to the electricity grid it falls even further, to only 3 times the cost of hydroelectric power. But as previously mentioned, the cost of solar systems has been falling every year and due to the valuation of environmental and social costs, solar systems tend to become economically competitive in the short term (SHAYANI; OLIVEIRA; CAMARGO, 2006).

The simplest way to analyse the cost is to compare energy sources through their implementation cost per unit of power, since excessive initial investments tend not to attract the attention of investors easily, especially when the interest rate is high. A study published in 2005 on the cost of installing photovoltaic systems analysed the price of 47 isolated systems ranging from 100 to 6600 W, from 1987 to 2004, indicating that these systems show a downward trend in price of approximately 1 U$/W per year, with costs varying between 7 and 10 U$/W.

Another study confirms that prices are decreasing year on year and shows that isolated systems tend to cost approximately twice as much when compared to grid-connected systems, because grid-connected systems don't need batteries and other adjunct components (Published by the International Energy Agency's Photovoltaic Power Systems Programme). In 2004, isolated systems of up to 1kW were worth more than twice as much as systems larger than 1kW, in this case grid-connected systems.

Considering that isolated photovoltaic systems typically cost around 13U$/W, it doesn't take much to realise that this system is surprisingly uneconomical and uncompetitive when compared to the cost of implementing other sources. But you can't just take the cost comparison into account. Another important consideration is the comparison of energy generation capacity over the course of a day. A fadado system with a non-intermittent source can generate energy for 24 hours a day, while a solar system with the same installed power can generate an average of 6 equivalent hours of nominal power over the course of the day, depending on its geographical location. Therefore, in order for the solar photovoltaic system to produce an equivalent amount of energy in one day, its power needs to be increased by 4 times, which increases its implementation cost to 52 US/WPICO. This way of presenting the figures proves to be a great ally in keeping fossil systems in increasing use, since solar energy is 50 times more expensive than small hydroelectric plants (SHAYANI; OLIVEIRA; CAMARGO, 2006).

Table 1 shows the typical implementation values for power plants.

Table 1: Typical implementation values for power plants

TYPE OF GENERATION	ANEEL IMPLEMENTATION COST (U$/W)	CESP/MT IMPLEMENTATION COST (U$/W)
DIESEL THERMOELECTRIC	0,40 à 0,50	0,35 à 0,50
GAS-FIRED POWER STATION	0,40 à 0,65	0,35 à 0,50
STEAM POWER STATION	0,80 à 1,00	-
THERMOELECTRIC COMBINED CYCLE	0,80 à 1,00	-
SMALL HYDROELECTRIC POWER STATIONS	1,00	-
WIND GENERATION	1,20 à 1,50	1,00
PHOTOVOLTAIC CELL	-	5,00 à 10,00

Source: (FERIOLI et al., 2014)

However, this account is inappropriate because it does not take into account the high cost of fuel for thermal power stations, an item that does not exist in solar systems. Furthermore, the cost of operation and maintenance is 5 times less in photovoltaic generation (Adapted from SHAYANI; OLIVEIRA; CAMARGO, 2006).Figure 4 shows the annual evolution of oil prices, in dollars per barrel, indicating an increase in recent years.

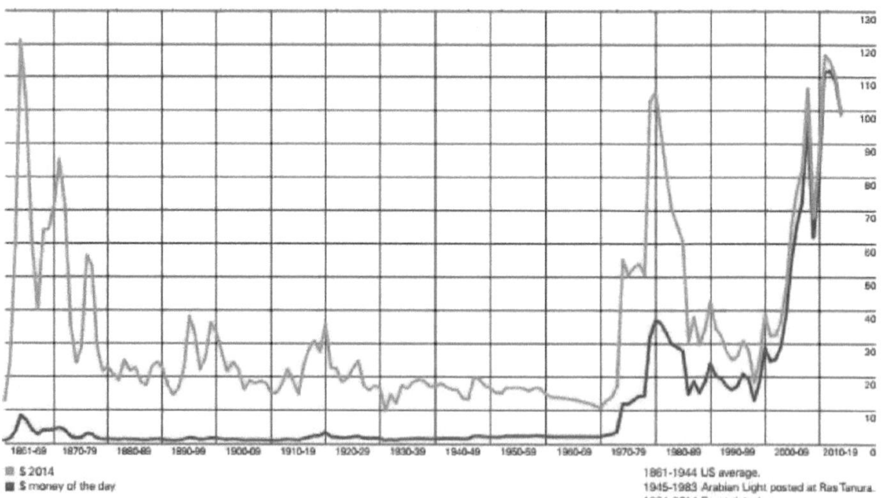

Figure 4: Oil prices 1861 -2014 (U$ dollars per barrel)
Source: (BP STATISTICAL REVIEW OF WORLD ENERGY, 2015)

A more accurate comparison of the real price difference between solar energy and other energy sources can be made using only technical criteria, not taking into account other factors whose valuation could be considered subjective. The following order is used:

1) Comparison using the price of the energy generated, rather than the installed power. Given that photovoltaic solar energy has negligible operating and maintenance costs, above all because it does not require fuel to operate and has no moving parts to undergo complex maintenance, the investment made in its installation is diluted over its entire useful life, corresponding to the energy generated;

2) Comparison with the value of energy from conventional sources that is paid by the consumer unit, after the transmission and distribution system, rather than the price charged by the generating plant.

The photovoltaic system used in distributed generation supplies energy directly to the consumer's home, and can take place on the roof of the consumer unit itself. In the case of conventional sources, the value that should be used as a reference is the energy charged by the distribution concessionaire for the residential class, which takes into account, among other costs:

- energy generated by the plant;
- transmission lines;
- distribution network;
- operation;
- maintenance;
- sector charges, especially the fossil fuel consumption account (CCC);
- various costs, such as the energy rationing that took place in 2001.

The energy delivered to consumers is several times more expensive than the plants' sale price. Table 2 shows the average energy prices at power plants that are negotiated at energy auctions.

Table 2: Average price of energy negotiated at auctions

Start of procurement	Average price (R$/MWh)	Energy trading event
2008	233,24	1° Reserve energy auction
2009	196,27	2° Reserve energy auction
2010	166,86	3° Reserve energy auction
2011	119,13	4° Reserve energy auction
2013	118,38	5° Reserve energy auction
2014	169,82	6th reserve energy auction

Source: Adapted from info Leilão (2014)

Table 3 shows the final cost claimed for energy from the distribution system in 2015 for the B1 conventional residential class at various distributors in Brazil. It can be seen that the average value is R$449.45 R$/MWh, which is more than 250% higher than the energy sold in the 2014 auction, making the price discrepancy with other sources, which have no costs associated with transmission and low distribution costs, more attractive.

Table 3: Amount of energy charged to residential consumer units in 2015, minus ICMS tax

Company/State	Residential energy tariff (R$/MWh)	Company/State	Residential energy tariff (R$/MWh)
Escelsa/ES	464,52	Celpe/PE	395,24
Cemig/MG	509,74	EPB/PB	418,17
Ampla/RJ	506,92	Cosern/RN	375,90
Bandeirante/SP	500,57	Coelce/CE	417,96
Eletrocar/RS	527,91	Cepisa/PI	439,87
IEnergia/SC	454,65	Cemar/MA	469,35
Cocel/PR	531,99	ETO/TO	462,03
Enersul/MS	464,70	Celpa/PA	525,39
EMT/MT	465,20	CEA/AP	272,62
Celg/GO	466,60	Boa Vista/RR	406,64
Coelba/BA	388,36	AME/AM	445,31
ESE/SE	409,35	Ceron/RO	472,35
Ceal/AL	439,11	Eletroacre/AC	463,27
Average value: 449.45 R$/MWh			

Source: Adapted from the Brazilian Association of Electricity Distributors (ABRADEE) (2016)

1.5 ADVANTAGES OF USING SOLAR ENERGY

The transformation of solar energy into electrical energy, through the photovoltaic phenomenon, takes place directly, in a process that is clean, silent and carried out at the place of consumption. This, combined with technological development, means that the production of electricity through photovoltaic generators is currently a technical and economic reality that is spreading around the world and also in Brazil, and is highly likely to reach new users.

Considering the country's territorial extension, the high solar radiation index, the air humidity levels, the dispersion characteristics of this population and the low electrification indicator in Brazil's rural areas, it can be stated that Brazil is destined to be an important user of this technology.

The benefits of the photovoltaic system, which add to the favour of its implementation, can easily be seen in the reliability and operational stability of this way of obtaining energy. This is in addition to the fact that this system is not aggressive to the environment and also allows for the electrification of remote areas, helping people to settle in their places of origin and instantly improving the living conditions of communities.

According to Silva (2010), after a period of crisis, there was a drop in the use of photovoltaic cells to produce electricity. As a result, the price of photovoltaic cells also fell. With movements to defend and preserve the environment, as well as the need to supply energy in rural areas, the use of photovoltaic cells for energy production has picked up significantly. Reducing the cost of photovoltaic systems is a particular challenge for the sector, which is being overcome through the development of new technologies.

One of the advantages of using photovoltaic systems to produce electricity is the fact that the consumer is also the energy producer, and in most cases the sole owner is responsible for the whole process. Another important advantage of using this system is that the energy source is free and available indefinitely (renewable energy source), since energy from the sun is available everywhere. Not to mention the advantage of having an energy supply in isolated locations and the fact that There is also local generation, which helps to preserve the environment, as the impacts are surprisingly low.

Photovoltaic systems are also highly reliable, as they have no moving parts and are not as complex as current energy generation methods, and are not seriously affected by external influences such as lightning and wind.

Another benefit of this system is the absence of disturbances and instability in the network. This, as well as the possibility of infinite expansion of the system, since the photovoltaic cells are arranged in modules, makes it possible to design and adapt the system to the intrinsic needs of each consumer; this lowers the cost of the initial investment, and provides reliability and energy availability.

One of the most recent and notable uses of photovoltaic energy is to provide energy for use in different recharging systems. There has been a lot of research into the use of the energy generated by the photovoltaic effect to power electrical and electronic equipment and even vehicle batteries (SILVA, 2010).

1.6 DISADVANTAGES OF USING SOLAR ENERGY

According to Silva (2010), however, even though it is a clean and possibly inexhaustible source of energy, solar energy, which is based on photovoltaic energy, and consequently the latter, does not produce yields comparable to other types of energy sources. It should be noted that although there is no considerable loss in the process of converting the electrical energy generated into chemical energy, the technology used to convert light energy into electrical energy still needs to evolve substantially in order to achieve a higher yield.

It is also worth emphasising that the amount produced and stored does not show sufficient autonomy to supply systems with a greater energy demand for a considerable period of time. This is due to the limitations of light-to-electricity conversion, as well as electricity-to-chemical conversion and storage capacity (this will be discussed in section 1.8.2). In this way, production, storage capacity and demand are limiting factors; the symptoms indicate that these limitations will soon be overcome with the development of the technologies employed (SILVA, 2010).

In addition to the disadvantages mentioned above, there are other factors that can be cited as disadvantages of this type of energy.

One of these is climatic influences (rain and snow). This factor causes variations in the quantities produced depending on what happens. In addition, it is worth remembering that at night, due to the absence of the sun, there is no production at all, which means that there must be means of storing the energy produced during the day in places where the solar panels are not connected to the energy transmission network.

Another factor that greatly influences the variation in the amounts of energy produced is latitude. Places at medium and high latitudes, such as Finland, Iceland, Norway, southern Argentina and Chile, among others, suffer sharp drops in production during the winter months due to the fall in the daily availability of solar energy (PORTAL ENERGIA, 2016).

In Finland, for example, for about three months, from mid-November to mid-January, the country is almost completely dark, considering that in some cities in the south of the country the time it takes for the sun to rise and set is only six hours a day, and worse, in some cities in the north of the country, sunlight is not present for more than an hour a day. This is due to the earth's translation movement. As the earth revolves around the sun, the sun's incidence changes, especially in countries at medium and high latitudes, as mentioned above and as can be seen in Figures 5 and 6.

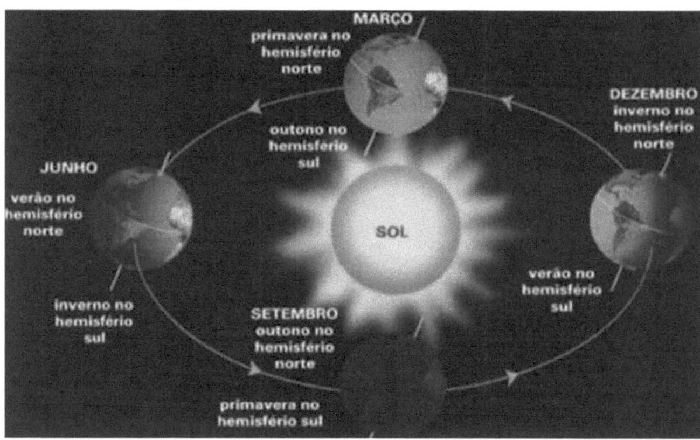

Figure 5: Earth's translation movement
Source: (Online history lessons, 2016)

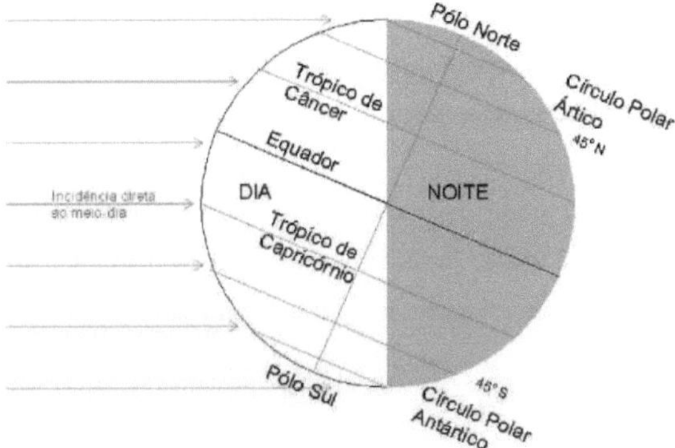

Figure 6: Solar radiation on the rotating earth.
Source: (Astrosurf, 2016)

When the earth's translational movement is configured as shown in figure 6, it can be seen that the sun's rays hit much harder between the Equator and the Antarctic Circle, located at the South Pole. In order to reach the North Pole, the sun's rays have a longer way to go, making the incidence much lower, causing the period of darkness in Finland mentioned above and also causing little solar radiation in other northern countries, making it unfeasible to use solar energy at this time of year in these northern countries.

And unlike some of the disadvantages mentioned here, which over time will no longer be seen as disadvantages due to the changes that have been taking place, such as storage capacity and demand, these last-mentioned disadvantages are permanent and there is nothing that can be done to reverse them.

1.7 RECOMMENDATIONS FOR THE MOST EFFICIENT USE OF PHOTOVOLTAIC SOLAR ENERGY

In order to achieve effective utilisation of the system, it is essential to observe some specific topics. Due to the high costs of photovoltaic systems, to the detriment of the amount of energy they generate, system sizing is a seriously critical element in a photovoltaic installation and will be discussed in section 3.4. It is therefore essential to adopt uncompromising sizing criteria, so that the system is adapted to the calculated loads and energy use estimates.

Oversizing the system generates a high increase in costs: it can make the project unviable. On the other hand, if the photovoltaic system is undersized, it will not be able to meet demand, which will discredit the technology.

The quality of the components and the skill with which they are installed are directly linked to the reliability of the system. Maintenance is a crucial element. Although photovoltaic systems do not require frequent maintenance, it is necessary to be aware of the fact that small problems can arise that will require the temporary paralysing of the energy supply for repairs. Preventive maintenance is therefore highly advisable.

Since the photovoltaic system has a limited productivity when compared to other forms of electricity acquisition currently known, it is essential that all the energy-consuming devices used are high-performance, in order to avoid wasting electricity. "In this way, fluorescent lamps, low-loss circuit breakers, high-efficiency inverters and the like are fundamental to the technical and economic viability of the system" (SILVA, 2010, p. 65).

It is worth remembering that, as mentioned above, the availability of solar energy for electricity production depends on a number of factors such as: the movement of translation, which causes the earth to position itself in different ways in relation to the sun throughout the year, as well as the position of the sun during the day, which directly affects the rate of solar incidence in the region analysed (SILVA, 2010).

1.8 PHOTOVOLTAIC SYSTEMS

"In just one hour, the sun pours more energy into the Earth than the global consumption of an entire year" (TEÓFILO et al., 2004, p. 2). A photovoltaic generation system can have numerous configurations, depending on its use. In most cases, it is made up of a set of complementary devices comprising a power conditioning and storage subsystem.

Since photovoltaic panels produce DC electricity, we also need an inverter in the subsystem. The inverter transforms DC into AC and can therefore power conventional electrical and electronic appliances. The storage system is made up of batteries, which store the electrical energy obtained from sunlight during the day, making it possible to run lamps and electrical appliances at night and/or during rainy periods. The conditioning subsystem also consists of a charge controller. This is installed between the modules and the batteries and coordinates the process of charging and discharging the batteries, preventing them from being overcharged or discharged beyond pre-established limits. In this way, the batteries' lifespan is extended (TEÓFILO; SOUZA; MESQUITA, 2004).

1.8.1 Photovoltaic panel

Photovoltaic panel, in other words photovoltaic module, is the name given to a system made up of photovoltaic cells that generate electricity.

The cells work on the basis of a phenomenon called the photoelectric effect, which occurs significantly in electrically conductive materials. This effect occurs when visible light with a reasonably high frequency (such as indigo, blue or violet) is allowed to reflect off a conductive metal. Under the incidence of the light, the metal begins to release electrons from its last layer, also known as the valence layer. The most highly regarded cells on the market today are made up of a combination of two semiconductor materials so that, according to a property of this category of material, they can use these electrons to produce an electric current. By means of an electric current, it is possible to capture electrical energy (SOUZA; SILVA; SILVA, 2010).

This photovoltaic effect was discovered by Edmond Becquerel in 1839. However, it wasn't until 1883 that the first photoelectric cells were built by Charles Fritts, who coated semiconductor selenium with a surprisingly thin layer of gold to form junctions (PORTAL ENERGIA, 2004, 2016b).

Nowadays, there are several photovoltaic panel technologies, three of which are the most widely available on the market. These are Monocrystalline silicon, polycrystalline silicon and amorphous silicon.

- Monocrystalline silicon

"The monocrystalline silicon cell has historically been the most widely used and commercialised direct converter of solar energy into electricity and the technology for its manufacture is a very well-developed basic process." (CRESESB, 2016 a, p. 4).

Under normal conditions, silicon can make four electronic bonds, depending on the number of electrons in its last layer (valence layer). The way in which these bonds are made will form various types of silicon. In the case of monocrystalline silicon, the bond is the one that appears to order the atoms the most, as shown in figure 7.

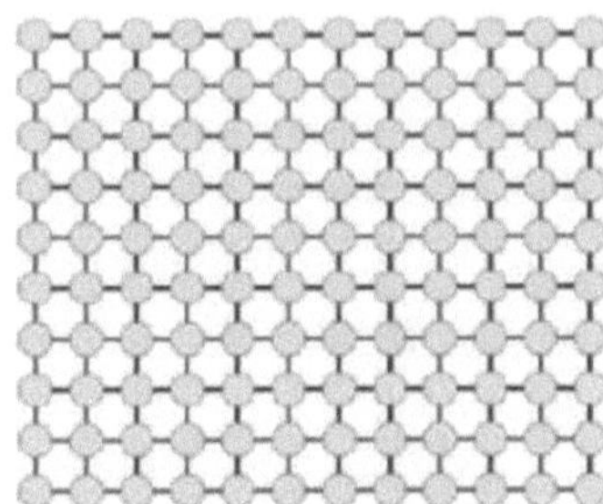

Figure 7: Monocrystalline silicon
Source: (SOUZA; SILVA; SILVA, 2010).

Boards using this type of silicon achieve yields of over 12%, where yield determines the amount of energy produced per energy incident on the panel. Figure 7 shows that the atoms in this silicon are highly ordered (SOUZA; SILVA; SILVA, 2010).

This type of photovoltaic cell represents the first generation. And although their performance is higher than other types of cells, the techniques used to produce them are complicated and expensive.

And due to the need to use materials in a very pure state and with a perfect crystal structure, a large amount of energy is required in the production process (PORTAL ENERGIA, 2004, 2016b).

- Polycrystalline silicon

Polycrystalline silicon is the middle ground, where its bonds are not as ordered as those of monocrystalline silicon, but also not as disordered as the bonds of amorphous silicon, as shown in figure 8.

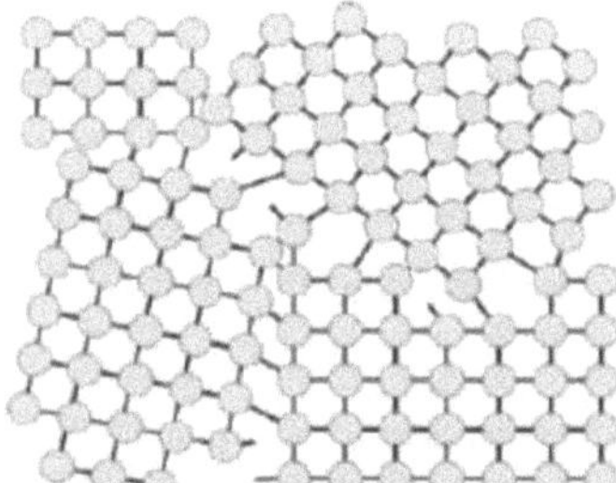

Figure 8: Polycrystalline silicon
Source: (SOUZA; SILVA; SILVA, 2010)

One advantage of this type of silicon is that it is simpler to produce than monocrystalline silicon, but its yield is also somewhat lower.

In this type of silicon there is even an ordering of the atoms, but this ordering is not as great as in monocrystalline silicon (SOUZA; SILVA; SILVA, 2010).

Another advantage of polycrystalline cells compared to monocrystalline cells is their lower production cost, as they require less energy to produce. But because of this, there is also a disadvantage, which is the lower electrical efficiency than monocrystalline cells. This performance attenuation is due to the imperfection of the crystal, thanks to the manufacturing system (PORTAL ENERGIA, 2016).

- Amorphous silicon

Also known as thin films, amorphous silicon is the silicon with the most disordered atoms of the three types mentioned here, as shown in figure 9.

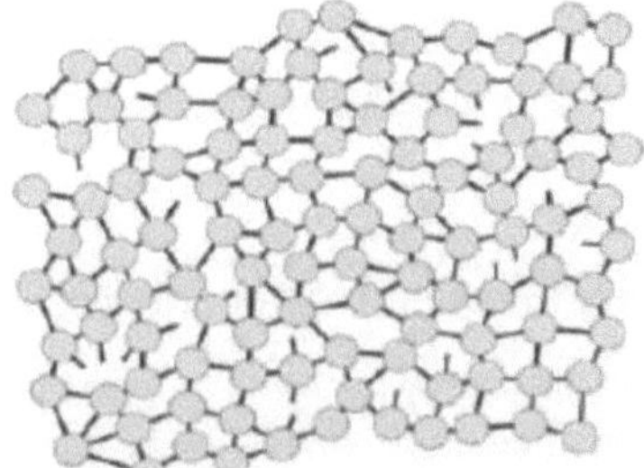

Figure 9: Amorphous silicon
Source: (SOUZA; SILVA; SILVA, 2010)

Compared to the other types, amorphous silicon is the technology that transforms the sun's rays into energy the least, with an efficiency of just 7%. However, it does have the advantage of being the one that uses the least silicon, and the thickness of one of these plates can be around 1 pm, as well as being very flexible.

As can be seen in figure 9, the ordering of amorphous silicon is less than that of monocrystalline and polycrystalline silicon.

Most silicon-based boards have a useful life of around twenty years, bearing in mind that this research analysed a board that works on the basis of monocrystalline silicon technology (SOUZA; SILVA; SILVA, 2010).

Figure 10 shows a cell of this type of silicon.

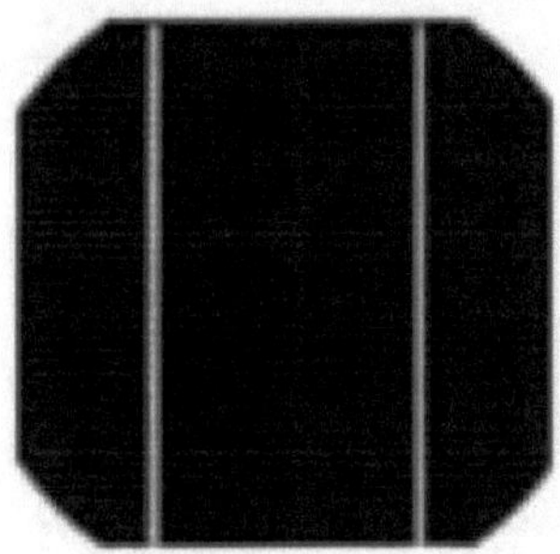

Figure 10: Amorphous silicon
Source: (SOUZA; SILVA; SILVA, 2010)
Generally, this type of cell is very dark and unicoloured, since it is made up of practically a single crystal of silicon, hence the term monocrystalline (SOUZA; SILVA; SILVA, 2010).

1.8.   2Battery

Electricity storage is a key issue when it comes to harnessing solar energy from autonomous systems, since energy production and consumption do not match up, either during the day or over the course of the year. The solar energy generated during the day is generally not used before night falls, which is why it needs to be stored. In addition, we must take into account periods of consecutive days without sun (PORTAL ENERGIA, 2004, 2016).

A battery is an electronic element that is capable, through a series of reactions, of converting chemical energy into electrical energy and vice versa. An ordinary battery can simply be charged or discharged using a set of wires.

In a photovoltaic system, the electrical energy produced during the day by the photovoltaic plate is transformed into chemical energy via the battery. At night, the chemical energy is transformed into electrical energy to supply the consumer.

The useful life of a stationary battery is approximately four years (SOUZA; SILVA; SILVA, 2010).

1.8.   3Load controller

When equipment is connected to a battery, the amount of electrical energy stored in it decreases over time. To prevent the battery from being completely discharged during longer periods without sunlight and with high consumption, it is appropriate to install a charge controller. This equipment monitors the battery's charge, preventing it from discharging completely and therefore increasing its useful life (SOUZA; SILVA; SILVA, 2010).

> The main functions assigned to battery charge regulators are as follows:- Ensure battery charging;
> - Avoid overcharging the battery;
> - Block the reverse current between the battery and the panel;
> - Preventing the occurrence of deep discharges (in the case of lead-acid batteries)
> (CARNEIRO, 2009, p. 22)

Figures 11 and 12 show, respectively, a charge controller that responds flexibly to the photovoltaic generation and storage system and a modest charge controller with protection against overloads and deep discharges, for independent systems of up to 60 W (PORTAL ENERGIA, 2004, 2016).

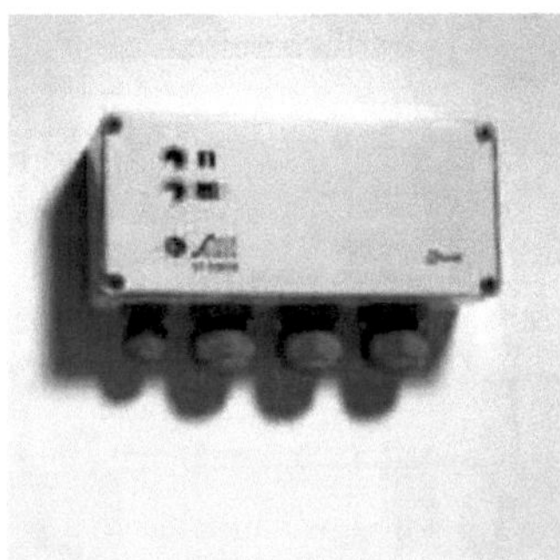

Figure 11: Charge controller for outdoor installations with charge status indicator

Source: (PORTAL ENERGIA, 2004, 2016)

Figure 12: NANO mini charge controller.
Source: (PORTAL ENERGIA, 2004, 2016).

1.8.4 Voltage inverter

A voltage inverter is an electronic instrument that works by varying the voltage and frequency of a given current.

The purpose of the inverter in a photovoltaic system is to change the energy produced by the plates and stored in the batteries from DC 12V to AC 220V, which is used to power the lamp on the pole. In most Brazilian cities, AC power is found in household sockets, while DC power supplies portable appliances, cars, computers (after their internal power supply), among others (SOUZA; SILVA; SILVA, 2010).

> Modern inverters have high efficiency, tracking of the maximum power point (SPMP) of the PV generator, safety measures for disconnecting from the grid in adverse conditions, anti-islanding mechanisms, measurement of electrical parameters, among other functions (ALMEIDA, 2012, p. 52).

To carry out the conversion responsible for the d.c./a.c. transformation, inverters use semiconductor elements such as static switches. These elements act in two states, on and off, so the output signal is modelled as a square wave. The output signal has a very powerful harmonic content, so filters are needed to achieve a pure sine wave.

The main semiconductor elements used are transistors and thyristors. Nowadays, transistors are the most widely used static switches. The transistors used are the MOSFET and the IGBT. Thyristors, on the other hand, are based on regenerative feedback from a PNPN junction, such as the SCR and GTO.

The method of filtering the harmonic elements requires large capacitors and inductors, thus limiting the efficiency of the inverter. One way to achieve output signals with low harmonic content and high power factor, without jeopardising efficiency, is to amplify the switching frequency of the switches and filter the output signal appropriately.

Another issue linked to power quality is balancing the electricity grid. By introducing current into a single phase, single-phase inverters generate an imbalance that can lead to instability.

Consequently, it is suggested to connect a maximum of 4.6 kW to a single phase and in the case of higher powers, multiple single-phase inverters can be used to ensure a symmetrical distribution between the three phases of the electricity grid, or three-phase inverters (ALMEIDA, 2012).

CHAPTER 2

ENERGY SCENARIO

2.1 WORLD ENERGY SCENARIO

2.1.1 Subtitle History of the period 1980/2014

Currently, the world's energy consumption is still based on non-renewable fuels such as coal, natural gas and oil.

Global warming due to the burning of these fossil fuels has led to a search for renewable energy sources and for technically and economically feasible ways of utilising them, so that they can be a sustainable alternative for humanity (WANDERLEY, 2013).

[a]The 21st Conference of the Parties to the United Nations Framework Convention on Climate Change and the 11th Meeting of the Parties to the Kyoto Protocol took place in Paris from 30 November to 11 December 2015.

COP21 sought to reach a new international agreement on climate, extending to all countries, with the aim of keeping global warming below 2°C. The UNFCCC was admitted in 1992 during the Earth Summit in Rio de Janeiro, but only came into force in March 1994.

This Framework Convention is a universal convention of principles, which admits the existence of man-made climate change and gives industrialised countries the lion's share of the responsibility for combating it.

The Conference of the Parties (COP) is made up of all the "States Parties" (countries participating in the convention) and is the deliberative body of the Convention. Each year they meet in a global session where decisions are taken to put into practice the targets for combating climate change.

On 12 December 2015 in Paris, a new global agreement aimed at combating the effects of climate change, such as curbing greenhouse gas emissions, was adopted by consensus.

"One of the objectives is to keep global warming 'well below 2°C', while also pursuing 'efforts to limit the temperature increase to 1.5°C above **pre-industrial** levels'" **(UNITED NATIONS IN BRAZIL, 2016, p. one).**

With regard to climate finance, the final text orders developed countries to invest 100 billion dollars every year in projects to combat climate change and adaptation in developing countries (UNITED NATIONS, 2016).

Figure 13 shows the world energy matrix in 1980 and 2007. Supply increased from 7,183 million tonnes of oil equivalent (toe) in 1980 to 12,029 million toe in 2007.

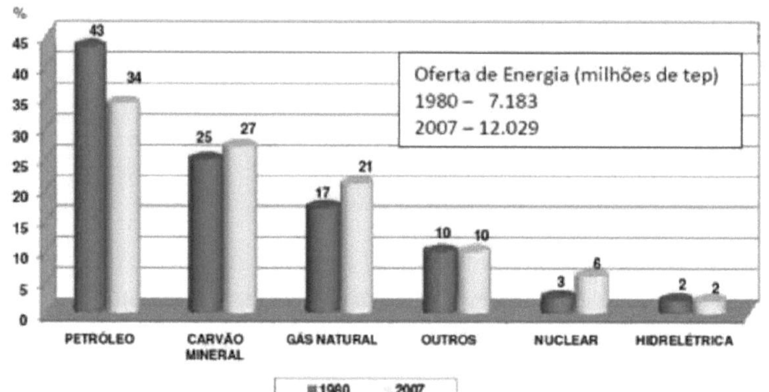

Figura 13:   World energy matrix
Source: (FILHO, 2009)

This figure shows that the world predominantly uses fossil fuels as a source of electricity generation, regardless of the year mentioned above. In 2007, oil and derivatives accounted for 34%, coal for 27% and natural gas for 21%, totalling 82%.

Despite the effort to reduce dependence on the "carbon age", during this 27-year period, consumption of fossil fuels increased worldwide. In total terms, the supply of these fuels rose from 6,133 million toe in 1980 to 9,792 million toe in 2007.

During this period, there was a modest improvement in the use of these fuels, with a switch from oil, which went from 43% to 34%, to natural gas, which went from 17% to 21%. From an environmental point of view, natural gas is the most favourable of the three, including in terms of carbon dioxide (CO_2) emissions, which is what made this small improvement possible (FILHO, 2009).

According to Filho (2009), the progress made by nuclear energy in the period between 1980 and 2007 helped to reduce the consumption of fossil fuels, especially oil and its derivatives, in the generation of electricity. Hydroelectric power stations, a renewable energy source that is only used to generate electricity, have kept their share at a constant 2%, showing that at global level this type of source is still little used when compared to some specific countries, such as Brazil for example. And this is where the Brazilian matrix differs from the global context, because here in Brazil hydroelectricity plays a fundamental role in supplying the country's energy demands.

26

During these 27 years, the world's energy matrix has not shown significant structural changes in terms of the use of primary energy sources. Since the industrial revolution, in order to meet their energy needs, the population has made intense use of fossil fuels. During the 19th century, preference was given to coal, in the following century (20th century) human society turned to oil and oil products, and in the current century humanity has turned to the three fossil fuels. As a result, the share of renewable energy sources in the current supply of the world's energy demands is only 14 per cent (FILHO, 2009).

2.1.2 The world's electricity matrix.

In terms of electricity specifically, the global subordination of fossil fuels is also high. Figure 14 shows the world's electricity matrix over the same period considered above, with the participation of the same sources mentioned in the world's energy matrix, as well as the supply of electricity over this period.

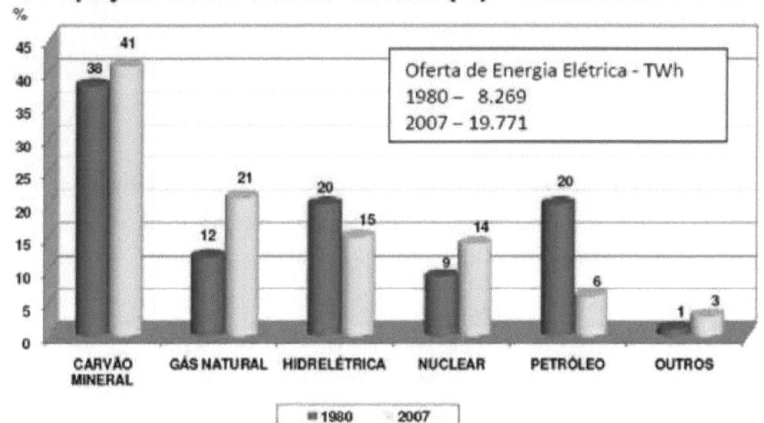

Figure 14: World electricity matrix
Source: (FILHO, 2009)

As can be seen in figure 14, supply grew from 8,269 TWh in 1980 to 19,771 TWh in 2007, resulting in an average annual rate significantly higher than the rate of total energy supply in the period, 3.2 per cent and 1.9 per cent respectively.

It can also be seen in figure 14 that the presence of fossil fuels in electricity generation between 1980 and 2007 fell slightly from 70 per cent to 68 per cent, and in total values grew from

5,779TWh to 13,469 TWh in the respective years, giving an average annual rate of development over the period of 3.1 per cent, which is very close to the rate of total electricity supply.

Today, coal is the largest source of electricity generation in the world, with a share of no less than 41%, followed by natural gas, hydroelectricity, nuclear power and finally oil and oil products. In the period from 1980 to 2007, the share of oil and oil products fell from 20 per cent to just 6 per cent, and nuclear increased from 9 per cent to 14 per cent. In global terms, renewable sources still have a small share, just 18 per cent, in electricity generation.

The international endeavour to reduce the use of oil and oil products in electricity generation has been successful, as can be seen in figure 14. This was only possible thanks to three main reasons:

1° - Disadvantageous economic competition between oil and oil products and other sources for generating electricity;

2 - The possibility of using these fuels in other more noble and appropriate areas other than the electricity sector;

3 - The availability of other energy sources that are more suitable for generating electricity, such as natural gas.

The widespread use of fossil fuels is evident due to four main aspects:

1 - The wide availability of resources, especially coal;

2nd - Favourable economic and environmental competitiveness (with the exception of oil and derivatives economically and CO_2 emissions and climate change) in relation to other primary energy sources.

3 - Advantageous technical and economic transport options, even over long distances;

4 - Appropriate and fully developed technology for its varied energy utilisation.

What's more, another important factor that justifies the widespread use of fossil fuels in the past, present and future is the simplicity of generating energy in the form of heat through them, with the weighted chemical reaction of carbon, which is abundant in them, with oxygen, the latter at zero cost, existing in abundance in the earth's atmosphere. The heat is used directly by consumers or simply transfigured into other forms of energy that are more appropriate for the user, such as the electricity generated in conventional thermal power stations (FILHO, 2009).

2.2 BRAZIL'S ENERGY SCENARIO.

2.2.1 History - Period 1980/2008

Let's now analyse the Brazilian matrix over practically the same period as the world matrix was analysed, differing only by the inclusion of 2008 in the Brazilian matrix.

Unlike the world energy matrix, Brazil's energy matrix from 1980 to 2008 was largely made up of renewable energy sources, mainly hydroelectricity and agro-energy (derived from charcoal, sugar cane and firewood).

Figure 15 shows the Brazilian energy matrix in 1980 and 2008, with the presence of the various energy sources. As we can see, the share of renewable energy sources in this time interval remained at 45 per cent, as a result of the energy policies adopted for hydroelectricity and agro-energy.

Unlike the world energy matrix, the national matrix has undergone considerable changes over the last 28 years in terms of the use of different energy sources. Thus, in addition to maintaining a high share of renewable energy sources, there has also been a significant reduction in the use of firewood and charcoal, falling from 27% to 11%, causing environmental benefits and increased yields. The share of sugar cane derivatives and hydroelectricity were favoured. The share of sugar cane derivatives doubled from 8% to 16% in the same period, while the share of hydroelectricity reached 14% in 2008, something unique in the world.

The main transformation in the national energy matrix during this period occurred in the case of fossil fuels. The share of oil and oil products actually fell from 48 per cent to 37 per cent, and was partly replaced by natural gas, which jumped from 1 per cent in 1980 to 10 per cent in 2008 (FILHO, 2009).

Table 4 will highlight the comparison between the world and Brazilian energy matrices, where we can see the big differences between them in terms of the share of energy sources. But in order to make a fairer comparison, I'm going to make the national energy matrix for 2007 available here, so that we can analyse the matrices over the same period to avoid small changes caused by economic growth, among other reasons, which varies every year.

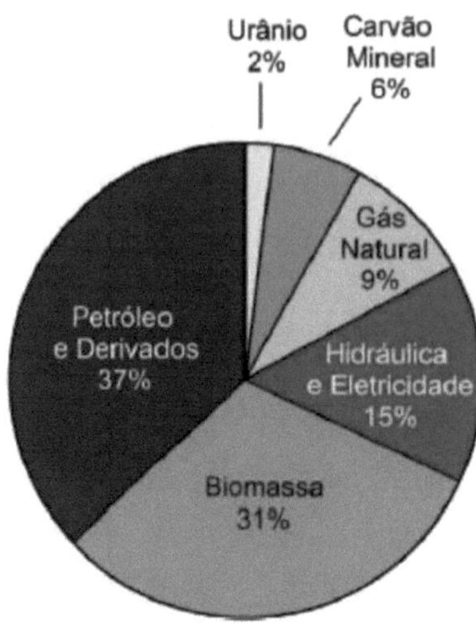

Figure 15: Brazil's energy matrix in 2007
Source: (VICHI; MANSOR, 2009)

The total amount of energy generated in Brazil in 2007 was 238.8 million toe, which represents 2% of all the energy generated in the world. A curious fact is the internal energy supply (IES) per capita, which in the same year was 1.29 toe/inhabitant, which is lower than the world average (1.8 toe), as well as being around 3.6 times lower than the average for countries in the Organisation for Economic Cooperation and Development (OECD), which is predominantly made up of rich countries (VICHI; MANSOR, 2009).

Table 4: Comparison of the share of different energy sources: Brazil and the world

Source	Brazil	World
Oil (%)	37	34
Biomass (%)	31	10
Hydraulics (%)	15	2
Mineral coal (%)	6	27
Natural Gas (%)	9	21
Uranium (%)	2	6
Millions of toe	226,1	12.029
Renewables (%)	45	12

Source: Adapted from VICHI; MANSOR (2009)

The table above shows the huge divergence between the shares of biomass and hydroelectric power between Brazil and the rest of the world, with a difference of 21 per cent and 13 per cent respectively.

2.2.   2Brazilian Electricity Matrix

Specifically in relation to electricity, figure 16 shows the Brazilian electricity matrix. This shows data for 2014, such as energy supply and the participation of the various energy sources.

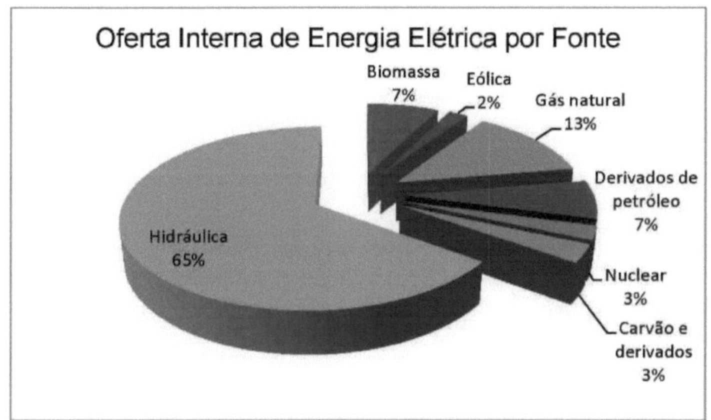

Figure 16: Brazilian electricity matrix
Source: BEN 2015

It can be seen from the figure above that the country's electricity sector is even more advantageous than the world's, with a large share of renewable energy sources, making a huge difference compared to the world scenario.

Thus, hydroelectricity is the country's main energy source, i.e. the energy source that produces the most electricity for Brazilians is renewable, accounting for 65 per cent of the total supply, given that almost all of the imported electricity comes from Paraguay's Itaipu binational hydroelectric plant. Fossil fuels, on the other hand, have a modest share in electricity generation in Brazil compared to the world figure (VICHI; MANSOR, 2009).

Another factor that can be seen in figure 16 is that solar energy doesn't even appear in the statistics. It can also be seen that BEN 2015 does not mention this type of source either, even though it is very promising for the country.

Solar energy is still so little used and mentioned in Brazil that only in January 2013 was the Brazilian Photovoltaic Solar Energy Association (ABSOLAR) created and only in 2014 was the first solar energy auction held in Brazil.

ABSOLAR was created precisely to try to break down these barriers that still exist in the solar energy sector in Brazil and to defend the interests of this industry.

Today, although it is still little used and talked about in Brazil, some incentives have been created for the population to join the system. For example:

o The state of Minas Gerais now exempts solar energy from the ICMS tax;

• You can now buy solar energy with Caixa's Construcard;

• BB is also providing a line of financing for those interested in adopting the system;

- The BNDES is financing photovoltaic panel factories to bring the technology to Brazil (PORTAL SOLAR, 2016).

CHAPTER 3

ON GRID PHOTOVOLTAIC SOLAR ENERGY PROJECT

3.1 SIZING THE PHOTOVOLTAIC SYSTEM

When designing a solar photovoltaic system, you need to know the meteorological characteristics of the location where the system will be installed and the consumption of the load to be supplied, because these two aspects directly influence the correct way to size the system.

With regard to meteorological information, this will be gathered from the website www.cresesb.cepel.br, which offers a programme called SunData, designed to calculate solar irradiation. The average insolation rates of the city of Venda Nova do Imigrante (VNI) in Espírito Santo will be taken into account, as this is the location with solarimetric data closest to the reality of Iúna, a city that was taken as a reference for the study, given that of the three nearby locations with solarimetric data cited on the website, VNI, although not the closest in distance, has the climate closest to that of Iúna (Adapted from SEGUEL, 2009).

The expected energy expenditure per day will be established taking into account an autonomous photovoltaic system of low electrical power, basically designed for lighting a middle-class home. All the elements of the system will be dimensioned according to the case study, with the aim of achieving a good match between the energy provided by the sun and the expected energy demand, a necessary requirement for any photovoltaic project (SEGUEL, 2009).

3.2 DETERMINING HOUSEHOLD CONSUMPTION

To calculate the size of the solar system, a model home was used. It should not be forgotten that each project must be analysed separately, paying attention to its consumption.

The model home is made up of 19 rooms where the electrical loads are distributed. Table 5 shows the number of electrical appliances used in the house, their power, the power used daily and the monthly power consumed.

Table 5: Household electricity consumption

Device	Power (W)	Utilisation (H/day)	Wh/day	KWh/month
Shower	16500	0,4*	6.487,5	194,6
Kitchen lamps	100	2	200	6
Bedroom	150	1	150	4,5

lamps				
Room lamps	150	3	450	13,5
Bathroom lamps	200	1	200	6
Light bulbs in other rooms	200	1	200	6
Television	110	3	330	9,9
Others	12565	0,5	6.282,5	188,5
Total	29975		14300	429

*Value rounded from 0.39 to 0.4, which equates to 23'59" or approximately 24'.

3.3 LOCAL SOLAR RADIATION LEVELS

Some locations may have a better yield than others; this varies according to the solar irradiation of the location. As already mentioned, data from the city of VNI will be used to analyse and insert the scenarios. However, we mustn't forget that although the data is from VNI, the project will be carried out in the city of Iúna. Figure 17 shows the solar irradiation in VNI.

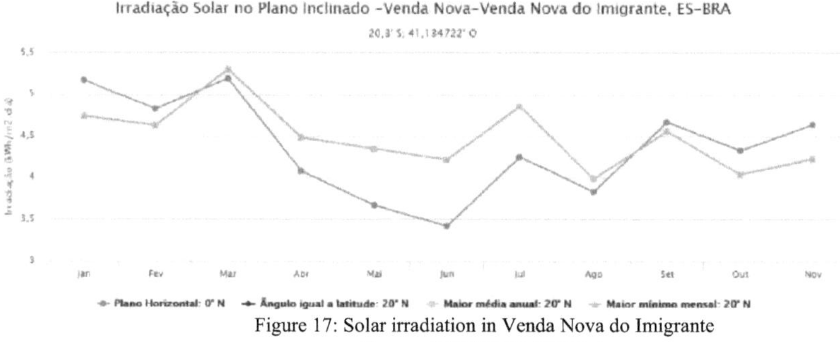

Figure 17: Solar irradiation in Venda Nova do Imigrante
Source: (CRESESB, 2016 b.)

Tables 6 and 7 will show what has already been seen in figure 17, but with greater precision in the monthly radiation values. Table 6 shows that the average annual radiation of the city of VNI in the horizontal plane is 4.31 kWh/m^2 and that June is the month with the lowest average solar radiation, with a monthly daily incidence of 3.42 kWh/m2.

Table 6: Monthly average daily solar radiation (kWh/m2)day for the city of VNI

Month	Jan	Feb	Sea	Apr	May	June	Jul	Aug	Set	Out	Nov	Ten	Average
Radiation	5,17	4,83	5,19	4,08	3,67	3,42	3,64	4,25	3,83	4,67	4,33	4,64	4,31

Source: Adapted from CRESESB (2016) b

Table 7: Monthly average daily radiation (kWh/m^2) for the city of VNI for the installation angle that provides the highest annual average (20° north)

Month	Jan	Feb	Sea	Apr	May	June	Jul	Aug	Set	Out	Nov	Ten	Average

												e
Radiation	4,74	4,63	5,30	4,48	4,35	4,22	4,43	4,86	3,99	4,56	4,04	4,23	4,49

Source: Adapted from CRESESB (2016) b.

Table 7 shows the solar radiation values for the plane inclined at 20° to the horizontal and pointing north. Under these conditions, according to the CRESESB website, the highest annual average is obtained, maximising the average of the month with the least insolation.

It should not be forgotten that the photovoltaic system needs to guarantee an energy supply every month of the year. Therefore, the lowest solar radiation index during the year should be used when sizing the photovoltaic system. Therefore, the design of a photovoltaic system for the city of Iúna should be carried out taking into account the incident radiation in September, which is 3.99kWh/m^2 (SEGUEL, 2009).

In addition to the monthly variation in radiation mentioned above, there is also daily variation in radiation.

During the course of the day there are certain hours when solar radiation is low or non-existent, while at other times solar radiation is intense, characterising a generation picture that is also variable, as we will see in figure 18. Taking into account all these variables, we can predict that at certain times of the day there will be electricity generation and at other times when the sun is non-existent there will be no production. In the same way that there is variation in generation, there is also variation in consumption and this occurs mainly at times when there is no energy generation, generally during the night. This makes it necessary to devise a scheme in which the energy generated during the day can be consumed at night. For this to happen, a compensation system is needed.

With the new Normative Resolution No. 482 of the National Electric Energy Agency (ANEEL), this credit compensation system has become mandatory for energy distributors. All the surplus renewable energy generated by consumers is injected into the electricity grid and returned to them in the form of energy credits. In this way, the distributor's electricity grid becomes a large energy storage bank for consumers, thus eliminating the need to buy batteries for this purpose.

Figure 18 shows that during the day the energy generated is consumed instantaneously and the kWh that exceeds consumption is transferred to the distributor in the form of energy credits. During the night, credits are offset, where the kWh generated and not used during the day compensates for the lack of generation during the night (PRÁTIL, 2016).

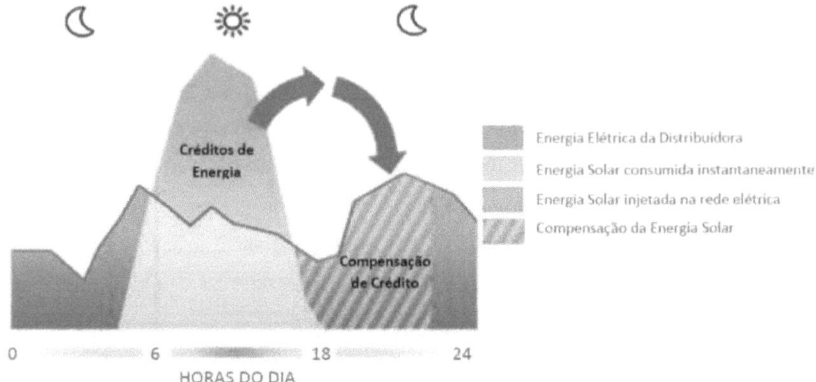

Figure 18: Daily variation in energy generation and compensation system.
Source: (PRÁTIL, 2016).

In addition to the compensation mentioned above, there is also another type of compensation that makes it possible to use the credits generated in the system at another location. For this to happen, all that is needed is for the electricity bill where the system was installed to be registered with the same CPF or CNPJ and the same energy distributor as the other location intended to receive the credit (PRÁTIL, 2016).

3.4 SYSTEM SIZING

Based on the energy consumption presented and the solar radiation of the location closest to where the system will be installed, as mentioned above, a system will be sized to meet a certain percentage of the model home's energy consumption. It is worth remembering that, just as there is variation in daily and monthly production and variation in daily consumption, there is also variation in monthly consumption. Table 8 shows the model home's monthly consumption during the holiday period, when there is more movement in the house, thus generating higher consumption. In order to calculate this more accurately, an average of the monthly consumption was made, based on the home's last four energy bills. The result of the average obtained will be considered to be the consumption for all the months of the year (BIG SOL, 2016).For the town of Iúna where the system will be installed, the initial proposal is shown in Table 8.

Table 8: System characteristics.

Photovoltaic system power	2.86 kWp
Area required by the system	18 m^2
Number of modules (boards)	11 units
Module inclination/orientation	20° / north

With these characteristics presented above, the system will generate an average of 318 kWh of electricity per month, which corresponds to 79 per cent of the average consumption of the home adopted, making it a partial energy generation system.

Installation with partial power can occur for reasons of technical limitations, such as the unavailability of sufficient area to install the complete photovoltaic panel or other limitations. However, once installed with partial power, the system can be expanded to full power.

CHAPTER 4

ECONOMIC VIABILITY ANALYSIS

In order to draw up an economic feasibility analysis for photovoltaic systems in general, it is always necessary to look at the local legislation. After all, because an economic analysis describes the project's financial gains, it is always necessary to be aware of the remuneration, provided for by law, applicable to the size of the photovoltaic plant under study.

There are various economic indicators for analysing a project. In the case of the project under study, the main indicators are payback, amortisation, NPV (Net Present Value) and IRR (Internal Rate of Return).

The work refers to the payback indicator. For a more detailed study of this indicator and the others that will not be cited here, we recommend the book Análise de Investimentos (MOTTA; CALÔBA, 2002) cited in the references.

In this section, we will show the payback for the proposed project for dimensioning the photovoltaic generation system to be realised in a model home. However, the calculation will not be detailed here, as this is not the objective.

With regard to the energy tariff, we will consider the latest annual update applied by Energias de Portugal (EDP) through the definition of the tariff adjustment granted by ANEEL, which was 2.29% from 07 August 2015 and will be valid until 06 August this year.

We can see a total increase in the variation of the energy tariff from 1995 to 2005 of 288.6%, which is why he considered that this figure remained so conservative. In our analysis, we will consider a period of 25 years, where the overall variation in this period is 172.18%, and in the first 10 years there was a variation of only 122.6%.

Table 9 details the cost of the equipment used, the installation of the system along with the freight charged for delivery and also the documentation required to approve the project in question. In this way, it is possible to calculate the total initial debt and how long it will take to pay back by adopting mini-generation (Adapted from MIRANDA, 2014).

Table 9: Total investment.

Item	Quantity	Description	Unit price.	Total price
01	11	Canadian Solar Panel 260 W	R$ 956,90	R$ 10.526
02	1	Grid-tie inverter 3.0KW ABB	R$ 5.990	R$ 5.990
03	3	Roof mounting bracket for 4 modules from 230W to 260W	R$ 860	R$ 2.580
04		System Documentation for Approval by the	R$ 1.000	R$ 1.000

		Dealer		
05	-	Other*	R$ 7.506	R$ 7.506
		Total = R$ 27,602.00		

*This includes: installation, electrical installation and safety materials and remote monitoring of generation performance via the internet.

4.1 PAYBACK

Payback is the precise period of time it takes to achieve a return on the investment made in any application. This expression is often used for electrical and energy efficiency projects to ascertain their economic viability.

Although it is a general study method, it is quite limited. This criterion does not take into account risk, monetary correction or financing. This method is nothing more than the value where the net profit equals the amount invested in the project analysed (MIRANDA, 2014).

Figure 19 calculates the payback on the initial investment of the model project used in this work. This calculation took into account the estimated NPV of R$65,800, the estimated IRR of 23%, energy inflation of 8% p.a. and the discount rate of 10% p.a. These figures were taken from the budget drawn up by the Big Sol company and from these we were able to calculate the discounted payback. This was calculated using an Excel spreadsheet available at[1] and adapted to the reality of the model project in this work.

By calculating the payback, we were able to arrive at the following figure, which shows the estimated payback time for the model system.

---

[1] https://eletricaefinancas.wordpress.com/2015/09/24/vpl-tir-e-payback-sistema-fotovoltaico/

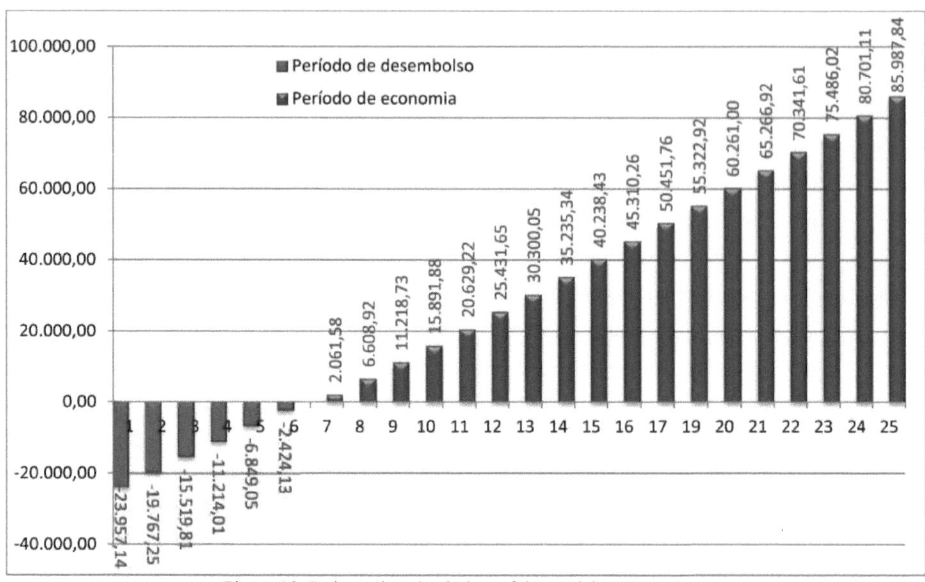

Figure 19: Estimated payback time of the model system

As can be seen in the figure above, the amount invested in the system will be returned in 7 years of generation and from then on all the generation will be transformed into savings, thus reaching a saving of R$ 85,987.84 in the twenty-fifth year. We can also see that 25 years of generation were considered. This is because the system guarantees generation performance for this period at 80% of nominal power.

4.2 SCENARIO BEFORE INSTALLING THE SYSTEM

Currently, the average electricity consumption of the home taken as the basis for this work is 400 kWh per month, which represents a cost of R$412.00 per month. Multiplying this figure by all the months of the year gives an annual cost of R$4,944.00 (BIG SOL, 2016).

Considering these values before the project was installed, i.e. when there was still no generation in the home, the following scenario was envisaged.

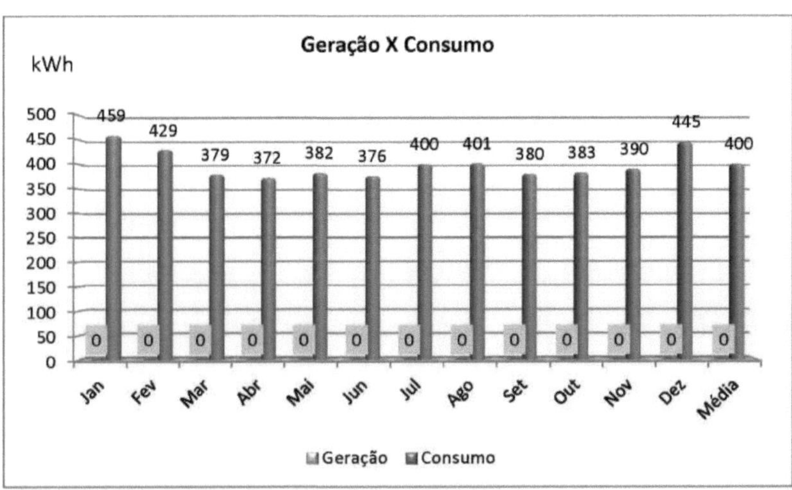

Figure 20: Generation X Consumption before system installation

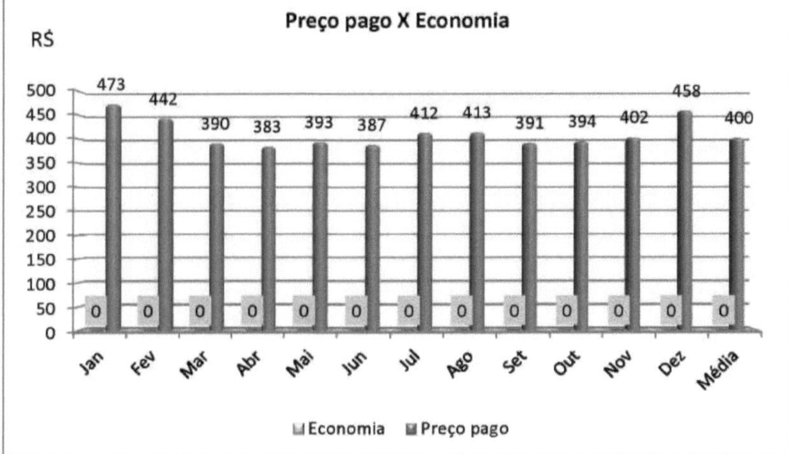

Figure 21: Price paid X Savings before system installation

The pictures above characterise the scenario before the project, where you can see the high consumption of the home, as well as the high price paid in the monthly electricity bills.

4.3   SCENARIO DURING SYSTEM INSTALLATION

With the installation of the system, the scenario changes completely.

Once installed, it will generate an average of 318kWh of electricity per month, resulting in an average 79% reduction in energy bills. This reduction will generate an average of R$ 325.00 in savings per month, totalling R$ 3,900.00 in savings in the first year. The following figures will describe this new scenario that takes place during the year the project is installed.

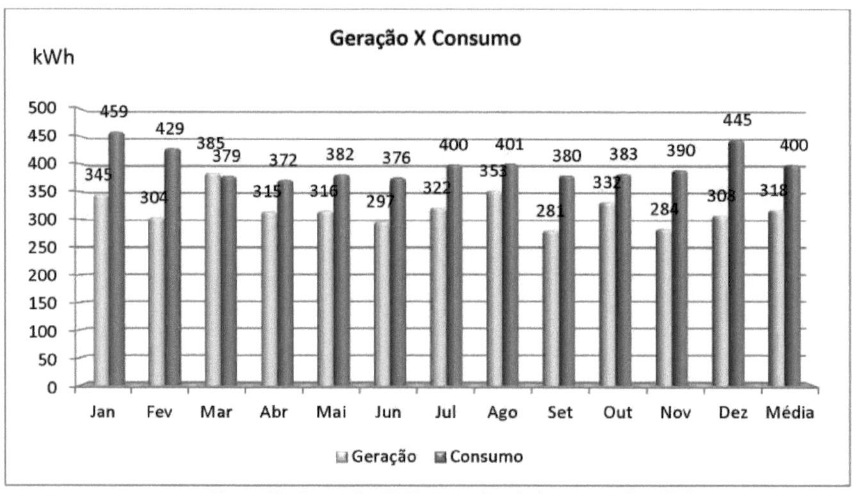

Figure 22: Generation X Consumption during system installation

Figure 23: Price paid X Savings during system installation

Figures 22 and 23 characterise the scenario during the project, where it can be seen that although the household's consumption remained high, as it was considered that consumption remained the same as before, the price paid in the monthly electricity bills decreased considerably, with some months seeing the bill go to zero, such as the month of March. This is because after the system is installed, it starts generating energy, and although consumption is still higher than generation, the amount generated monthly by the system reduces the amount supplied by the distributor, making this huge difference in the final cost consumed by the household.

As can be seen in Figures 22 and 23, the energy bill, which used to be an average of R$400.00 per month, now averages R$86.50, showing how viable the installation of the photovoltaic system is.

This cost reduction in the energy bill is equivalent to a return on the capital invested in acquiring the system of around 1.2 per cent per month or 14 per cent per year (BIG SOL, 2016).

If we compare this return with the return on savings, which today is around 0.65% per month or 7.8% per year (FINANCEONE, 2016), we can see that the return on the capital invested in the system is almost double the return on savings today. The following figure shows a graph that considers an initial value, which is the same as the cost of installing the project, and its respective return when it is invested in the acquisition of the system and in savings for a period of 10 years, also taking into account a saving of R$ 325.00 in the first month for both types of investment, the same saving that the system generates in the first month.

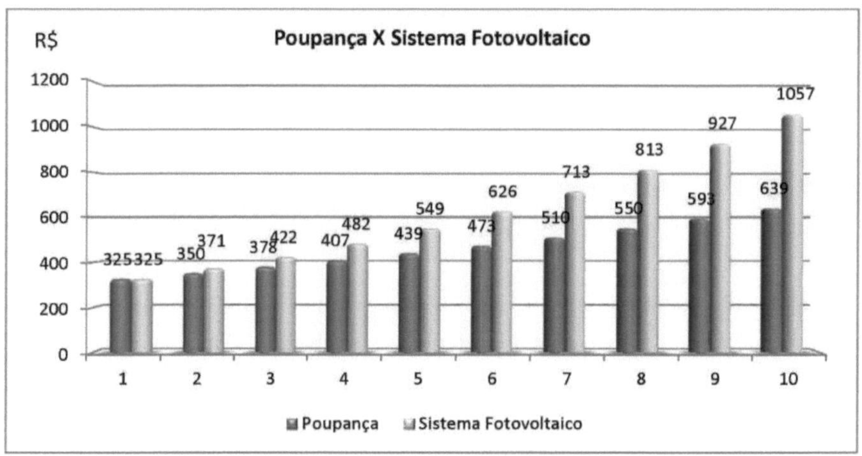

Figure 24: Investment in savings vs. investment in photovoltaic system

As can be seen in figure 24, it is much more profitable to invest in the photovoltaic system than in savings. This is without taking into account that if the money is invested in savings instead of in the installation of the project, the energy bill will continue to be high and the return from savings will be largely used to pay the monthly energy bill, which would not be the case if it were invested in the photovoltaic system.

4.4 PROJECTION ON A LARGER SCALE.

In Brazil, total electricity generation totalled 590.5 TWh in 2014. Of this total generated, 84.1 per cent came from public service power stations, the main source of generation being hydroelectric power, although there was a 4.5 per cent reduction in this source compared to the previous year. The other 15.9% of total production came from self-producers (APE), totalling 94 TWh when considering all the sources used. Of this total, 52.2 TWh is generated and consumed at the generating facility itself (BEN, 2015).

According to the Energy Research Company (EPE), in 2014 electricity consumption in the residential sector was 132,049 GWh, which is equivalent to 132 TWh (EPE, 2016).

Portal Brasil reported in 2011 that the number of residential units consuming energy in Brazil is 58.3 million (PORTAL BRASIL, 2016).

Considering all the data cited in this section and not taking into account the growth in these figures that occurred after the dates cited, assuming that 50 per cent of homes in Brazil were to install a photovoltaic system like the one presented in this paper.

If 50 per cent of Brazilian households (29.15 million) were to use photovoltaic systems, generating around 4,796 KWh per year for each household, which is the amount generated by the project presented in this paper, there would be a total generation of 139.8 TWh per year, which would exceed the residential electricity consumption considered here, which is 132 TWh, and there would still be a surplus to be consumed in other sectors.

According to BEN (2015), hydroelectric power produced 373,439 GWh in 2014, which is equivalent to 373.4 TWh. According to PEREIRA (2014), 3,600 litres of water are needed to generate 1 kWh of energy. This means that in 2014, 1,344.24 trillion cubic metres (m^3) of water were used to generate electricity, which is equivalent to 488,814,545.5 Olympic swimming pools, whose measurements are 50m x 25m x 2.20m, giving a total of 2,750,000 litres or 2,750m3 (Adapted from PEREIRA, 2014).

Therefore, if the photovoltaic system really took off, this amount could be used to replace part of the amount generated by hydroelectric power stations, for example, which, despite being a renewable source, need a lot of water, as we saw in the paragraph above, to generate 1kWh and, by reducing dependence on hydroelectric power stations, would also reduce water consumption, thus saving 503.28 trillion m3 of water.

In 2014, Brazil went through long periods without rain, causing a massive nationwide drought. With small measures taken on a large scale, such as the one we are proposing here, we can see that they would alleviate what many families are experiencing today, which is a lack of water. The use of solar energy to replace hydraulic energy is of great importance for the preservation of water in our territory (Adapted from PEREIRA, 2014).

And of course you can use solar energy to replace not only hydroelectric plants, but also all other types of energy sources, as is the case in Tokelau, a country that belongs to New Zealand and consists of three small islands in the South Pacific Ocean.

This is one of the most remote countries in the world, and it recently became the first country to be 100 per cent powered by solar energy. They have a 1MW system, which may be one of the largest isolated systems in the world, generating an amount that exceeds the needs of its local inhabitants.

Until then, the country had depended on diesel generators for its energy needs, but this change not only eliminated its energy dependency, but also made the country a world reference for the green movement.

These diesel generators used to use around 200 litres of fuel every day. Today, the country has not only completely eliminated its dependence on this fuel, but has also used the funds that were previously used to buy diesel to pay for the investment made in the photovoltaic system. What's more, with the additional funds caused by the savings that this investment has made possible, the country can also count on investing in the state's education and health areas (SOLAR PHOTOVOLTAIC PANELS, 2016).

This shows just how viable the use of solar energy is and how using it would bring about a perfect scenario if we look at all the water savings that would be generated and the reduction in greenhouse gases.

CHAPTER 5

CONCLUSION

Solar energy is becoming an increasingly important source of energy on the world stage. Because it is a clean and renewable energy source, solar energy is a good option to replace other types of sources that are not as viable as this one, especially those sources that generate energy by burning fossil fuels, thus causing even more environmental problems with the high CO_2 emissions from these sources.

Through the model project presented in this work, it can be seen that the use of solar energy brings another great benefit in addition to the non-emission of CO_2 and other benefits arising from the fact that it is a clean and renewable energy, which is technical and economic viability.

This work has presented some data showing how to harness solar radiation to generate solar energy, which is through the photovoltaic system and how profitable this investment is.

As previously mentioned, one of the great benefits of solar energy is the financial savings. In this work, graphs have also been presented to prove these savings, showing that the return on the investment made in installing the photovoltaic system is 6.5 years for the model presented here, and it can be said that the return was obtained in the short term, given that the system's guarantee is 25 years.

According to the data presented in Chapter 4, it can be seen that all the money invested in installing the project will be returned in 6.5 years, as previously stated, in addition to the fact that it already saves a total of R\$3,900.00 in the first year, thus reaching a saving of R\$85,987.84 in the twenty-fifth year.

Therefore, a family with a system identical to this one installed and with the same consumption as the model home, could pay off the entire investment made in installing the system in 10 years and still make a total saving of R\$15,891.88.

It can also be seen that if 50 per cent of the Brazilian population installed a system similar to this one in their homes, the total amount generated over the course of the year would be 139.8 TWh, which would supply the entire energy consumption of the Brazilian residential class, which is 132

TWh, with leftovers to be used in other sectors. Not to mention the water savings that would be generated, which could be used for other purposes.

ANNEX A - Photovoltaic solar energy systems.

A photovoltaic solar energy system is a system designed to produce electricity using solar radiation. There are two basic types of photovoltaic systems:

1° - Off-grid systems, which are systems isolated from the grid;

2° - On-grid or Grid-tie systems, which are systems connected to the electricity grid.

In this work we are using the on-grid system.

An on-grid photovoltaic solar energy system is made up of two basic elements, as shown in the figure below.

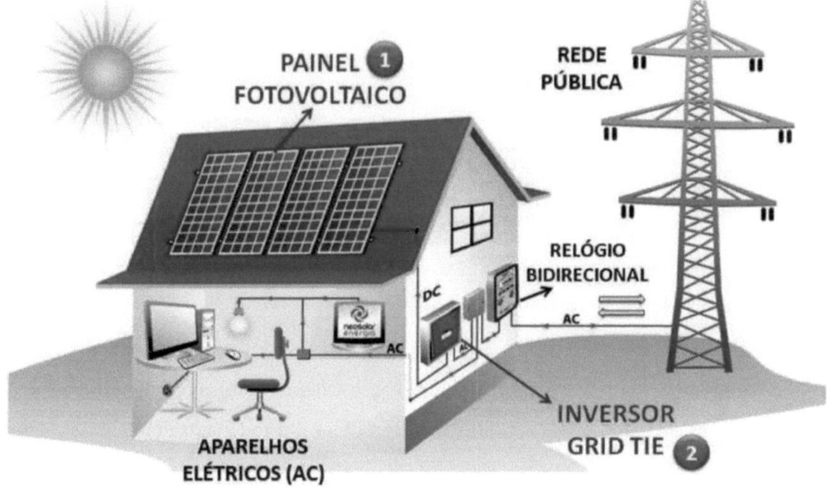

Figure 25: Preparing the installation site for the solar panels
Source: (NEOSOLAR, 2016)

Photovoltaic panel - The photovoltaic panel can be made up of one or more plates that are sized according to energy needs. The photovoltaic panel is responsible for transforming solar energy into electricity and using an analogy with the human body, we can consider that in the photovoltaic system the panel plays the role of the heart, "pumping" energy to the rest of the system.

Inverters - Inverters are responsible for transforming the 12 V AC of the batteries into 110 or 220 V AC or other desired voltages. In on-grid systems, inverters are also responsible for synchronising with the electricity grid. Using, once again, the analogy with the human body, we can consider the inverters to be the brain of the system (NEOSOLAR, 2016 a).

The steps that must be taken to install the on-grid system are now described in simplified form.

Firstly, we need to prepare the site where the solar panels will be installed. At this stage, the installation team climbs onto the roof of the house where the system is to be installed and plots where each solar panel will be located.

After tracing where each solar panel will be placed on the roof, it's time to install the panel "brackets". On clay roofs, which is the case with the model home, the tiles are removed in the correct places and the "brackets" are attached at these points, providing the base for fixing the system. When the roofs are made of metal, the installation process is even simpler. In these cases, the support is attached via the metal tile itself, providing security and protection against infiltration.

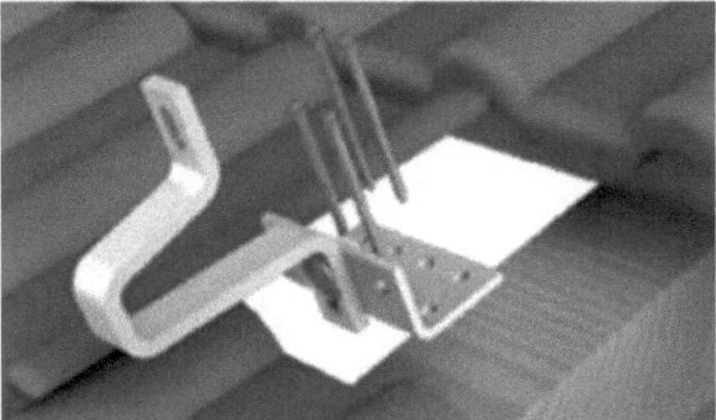

Figure 26: Preparing the installation site for the solar panels
Source: (SOLAR, 2016 a)

Then the "rails" are installed. The rails are the places where the solar panels will be fixed. They are built to fit precisely into the brackets and provide a precise place to attach the solar panels. All

these fixing structures are prefabricated and are generally made of aluminium (SOLAR, 2016 a).

In a typical structure, the solar panels will be placed on two parallel rails arranged at about % of the length or width of the panel. Depending on the slope of the installation plan, it may be necessary to use an angled structure, which is the case with horizontal slabs or roofs. In this situation, greater attention needs to be paid to dimensioning the wind loads.

It is recommended that solar panels be placed 10 to 15 cm away from the installation surface when they are positioned parallel to the surface. This distance makes it easier to make the electrical connections, but above all, it allows the solar panels to cool down and improves the efficiency of the photovoltaic cells (NEOSOLAR, 2016 b).

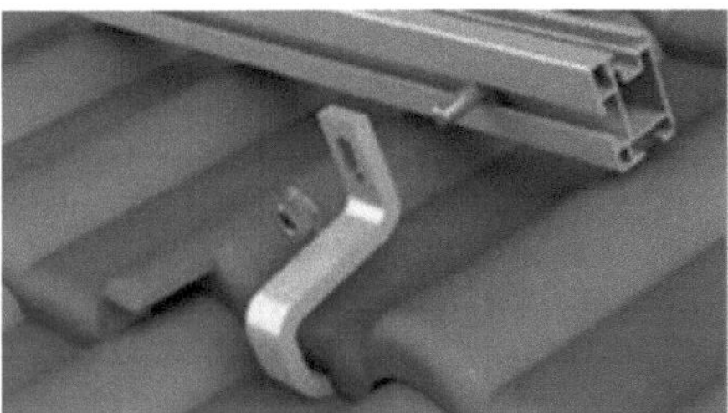
Figure 27: Installation of the rails on which the panels will be fixed
Source: (SOLAR, 2016 a)

After installing the rails and making sure that they are securely fastened, it's time to install the solar panels on the rails and connect the cables.

The aluminium structure of the panel has holes for fixing, as can be seen in the picture above, but the most common way is to fix by pressure, using clamps that hold one or two plates (panels) together, pressing the aluminium structure to the profiles. To do this, four clamps are used per panel, two clamps for each profile.

A minimum distance of 2 cm is recommended between each panel, favouring cooling and allowing room for expansion. Another interesting thing to do at this stage is to leave the cables and connectors out, to make it easier to connect them (NEOSOLAR, 2016 b).

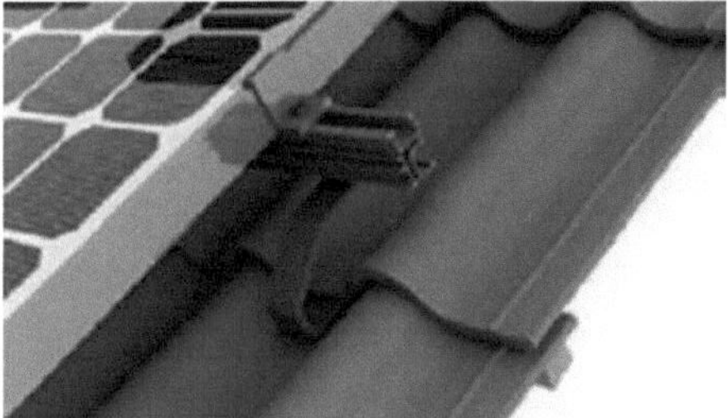
Figure 28: Installing the solar panels on the rails and connecting the cables
Source: (SOLAR, 2016 a)

Once this is done, we move on to connecting the solar panels to the solar inverter and installing the inverter in the house's electricity grid. After installing the inverter and connecting it to the grid, the photovoltaic system is already generating electricity (SOLAR, 2016 a).

The inverters can be placed on a wall or in a specially built structure, which is ventilated and easy to maintain and control.

The connection box with SPD and A.C. and C.C. switches must be placed close to the inverter. In certain countries, the disconnection switch must be unobstructed by the power distributor operator, making it easier for the operator to switch it off if necessary, as an additional safety measure. Here in Brazil, these connection details will be established by each distributor by the end of this year.

The next step is to connect the series of panels.

At this stage, connectors suitable for solar energy are very useful, although they can often seem complicated. For a series connection, simply connect them, male and female, following the series planning prepared for the photovoltaic array.

Because we are talking about DC, there will always be some risk of arcing. Connectors are used to contain this risk, which in no way reduces the need for attention. It is advisable to cover the panels during the electrical installation so that they do not produce current. Furthermore, the use of personal protective equipment (PPE) is essential.

After connecting the panels to each other, the remaining lengths of cable need to be coiled up and attached to the structure below the panels with a clip that can be made of plastic or nylon in order to reduce weather degradation (NEOSOLAR, 2016 b).

Figure 29: Inverter fixed to a wall
Source: (SOLAR, 2016 b)

Once the series of panels have been connected, they need to be connected to the inverters. To do this, you'll need to run an extension cable to the place where the inverters are fixed. Specifically, solar cables and connectors need to be used. If the design calls for two or more arrays to be connected in parallel, this needs to be done in the junction box.

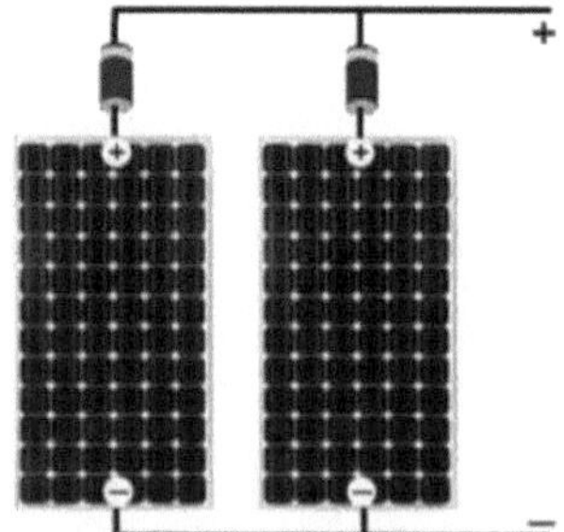

Figure 30: Parallel connection of two series
Source: (MPPTSOLAR, 2016)

Once the series of panels have been connected to the inverters, it's time to connect the inverter to the grid. The connection between the inverter and the grid is made with common cables used for AC. These cables must be connected directly to the busbars in the electrical panel, following the electrical design and phases.

And last but not least, the microinverters are placed in the very structure that holds the panels. Each panel is connected to its microinverter and the microinverters to each other with the connectors already in place. Because the power is supplied by AC, the cables used are ordinary cables, only with ultraviolet (UV) protection (NEOSOLAR, 2016b).

- Commissioning and connection.

Commissioning is the method that ensures that systems and components are all in compliance with operational obligations and requirements.

Commissioning can be separated into three phases:

1. Visual check: In this phase, the presence and quantity of calculated elements and the position and inclination of the photovoltaic panels are checked.
2. Mechanical check: Here we look at the fastening structure, electrical connections, earthing and connectors, sometimes on the A.C. side and sometimes on the C.C. side.
3. Operation: Drive the system from the panels to the grid. Check the status of the inverters and check the voltage and current on the A.C. and C.C. sides against the expected values (NEOSOLAR, 2016 b).

CHAPTER 6

BIBLIOGRAPHICAL REFERENCES

ABRADEE, Energy tariffs. Available at: <http://www.abradee.com.br/setor-de- distribuicao/tarifas-de-energia/tarifas-de-energia>. Accessed on 26 April 2016.

ACTION4ENERGY, United Nations Development Programme, Could you live without light? Available at: <http://www.action4energy.org/pt/>. Accessed on 21 December 2015

ALMEIDA, M. P. Qualification of grid-connected photovoltaic systems. Postgraduate Programme (Energy). São Paulo - SP: USP, 2012. 171 p.

BEN 2015. Final report - National Energy Balance, 2015. 291 p.

BIG SOL, quotation sent by the company Big Sol by email. Accessed on 18 April 2016.

BIGGI, R. R. The use of sunlight as a source of electricity through a photovoltaic system - SF. Postgraduate programme (Alternative forms of energy). Lavras - MG: UFLA, 2013. 41 p.

BP STATISTICAL REVIEW OF WORLD ENERGY, 2015. 64ª . ed.

CARNEIRO, J. Dimensioning photovoltaic systems (grid-connected and stand-alone systems). Master's thesis. Azurém: University of Minho, 2009.

CCEE, info Leilão, In: Reserve Energy Auction, 6th, 2014, info Leilão, CCEE, 31 October 2014, p. 9-11.

CRESESB, Solar potential - SunData, Geographical coordinates. Available at: <http://www.cresesb.cepel.br/index.php?section=sundata>. Accessed on 12 April 2016 b.

CRESESB, Tutorial on photovoltaic solar energy, Types of cells. Available at: <http://cresesb.cepel.br/index.php?section=com_content&lang=pt&cid=321>. Accessed on 7 April 2016 a.

DIETRICH, E. The movements of the earth. Available at: <http://aulasonlinedehistoria.blogspot.com.br/2015/08/os-movimentos-da-terra-os-movements.html>. Accessed on 6 April 2016.

DINIZ, F. B. The effects of latitude on the length of days and nights. How long are the days and nights? Available at: <http://www.astrosurf.com/skyscapes/disc/latitude/latitude.html>. Accessed on 6 April 2016.

EPE, Annual electricity consumption by class (national) - 1995-2014. Available at: <http://www.epe.gov.br/mercado/Paginas/Consumonacionaldeenergiael%C3%A9tric aporclasse%E2%80%931995-2009.aspx>. Accessed on 26 April 2016.

FERIOLI, K. C. O; VILHENA, A. L.; AGUIAR, M. A. S.; ARRIFANO, R. C. D; CORRÊA, F. Design of an Isolated Photovoltaic System (OFF-GRID) for a Home. Electrical Engineering Magazine, n. 2, 2014.

FILHO, A. V. O Brasil no contexto energético mundial. v. 6. São Paulo, SP: NAIPPE/USP, 2009.

23 p.

FINANCEONE, Savings. Available at: <http://financeone.com.br/>. Accessed on 21 April 2016.

MIRANDA, A. B. C. M. Economic feasibility analysis of a grid-connected photovoltaic system. Graduation (Electrical Engineering), Rio de Janeiro - RJ: UFRJ, 2014. 85 p.

MOTTA, R. da R.; CALÔBA, G. M. Análise de Investimentos: Decision Making in Industrial Projects. São Paulo: Atlas, 2002.

MPPTSOLAR, Parallel connection of more solar panels. Available at: <http://www.mpptsolar.com/pt/paineis-solares-em-paralelo.html>. Accessed on 16 April 2016.

UNITED NATIONS IN BRAZIL (ONUBR), United Nations Conference on Climate Change COP21/CMP11. Available at: <https://nacoesunidas.org/cop21/>. Accessed on 9 April 2016.

NEOSOLAR, Installation of the grid-tie system. Available at: <http://www.neosolar.com.br/aprenda/saiba-mais/sistemas-conectados-grid- tie/instalacao-do-sistema-grid-tie>. Accessed on 16 April 2016 b.

NEOSOLAR, Photovoltaic solar energy systems and their components. Available at: <http://www.neosolar.com.br/aprenda/saiba-mais/sistemas-de-energia-solar- fotovoltaica-e-sus-componentes>. Accessed on 16 April 2016 a.

Photovoltaic solar panels, Germany already produces half of its energy using solar panels. Available at: <http://www.paineissolaresfotovoltaicos.com/paineis-fotovoltaicos/a-alemanha-utiliza- paineis-solares/>. Accessed on 16 March 2016 a.

Photovoltaic solar panels, Tokelau the country 100 per cent driven by photovoltaic panels. Available at: <http://www.paineissolaresfotovoltaicos.com/paineis- fotovoltaicos/tokelau-o-pais-100-impulsionarado-a-paineis-fotovoltaicos/>. Accessed on 26 March 2016 b.

PEREIRA, I. The impact of reducing electricity waste on water savings. Available at:

<http://www.administradores.com.br/artigos/cotidiano/o-impacto-da-diminuicao-do- desperddicio-de-energia-eletrica-para-economia-de-agua/82151/>. Accessed on: 26 April 2016.

PORTAL BRASIL, Brazil has 68.6 million energy consumer units including homes, industries and public bodies. Available at: <http://www.brasil.gov.br/infraestrutura/2011/02/brasil-tem-68-6-milhoes-de- unidades-consumidoras-de-energia>. Accessed on 26 April 2016.

Portal energia Energias renováveis, Main types of photovoltaic cells that make up solar panels. Available at: <http://www.portal- energia.com/principais-tipos-de-cellulas-fotovoltaicas-constituintes-de-paineis- solares/>. Accessed on 7 April 2016.

Portal energia Energias renováveis, Advantages and disadvantages of solar energy. Available at: <http://www.portal-energia.com/vantagens-e-desvantagens-da- energia-solar/>. Accessed on 6 April 2016.

PORTAL ENERGIA. Technical guide Photovoltaic energy - Manual on technologies, design and installation, 2004. 368 p.

PORTAL SOLAR, Solar energy in Brazil, Known initiatives in Brazil.
Available at: <http://www.portalsolar.com.br/energia-solar-no-brasil.html>. Accessed on 17 June 2016.

PRÁTIL, quotation sent by the company Prátil by email. Accessed on 18 April 2016.

RATHMANN, R; BENEDETTI, O; PLÁ, J. A.; PADULA, A. D. Biodiesel: A strategic alternative in the Brazilian energy matrix? II Seminar on Business Management, Curitiba, UNIFAE, V. 1, p. 3-4, 2005.

REIS, L. B. dos. Electricity generation. 2ª. ed. Barueri, SP: Manoele, 2011. 460 p. Available at: <http://estacio.bv3.digitalpages.com.br/users/publications/9788520430392/pages/_1>. Accessed on: 21 December 2015.

REIS, L. B. dos. Electricity generation. 2nd ed. Barueri, SP: Manoele, 2011.
460 p. Available at:
<http://estacio.bv3.digitalpages.com.br/users/publications/9788520430392/pages/_1>. Accessed on: 15 March 2016.

REIS, L. B. dos. Electricity generation. 2nd ed. Barueri, SP: Manoele, 2011.
460 p. Available at:
<http://estacio.bv3.digitalpages.com.br/users/publications/9788520430392/pages/_1>. Accessed on: 16 March 2016.

REIS, L. B. dos; FADIGAS, E. A. A; CARVALHO, C. E. Energia, recursos naturais e a prática do desenvolvimento sustentável. 1st ed. Barueri, SP: Manoele, 2005. 418 p. Available at:

<http://estacio.bv3.digitalpages.com.br/users/publications/9788520420805/pages/_1>. Accessed on: 13 March 2016.

SEGUEL, J. I. L. Design of an autonomous photovoltaic energy supply system using MPPT technique and digital control. Postgraduate Programme (Electrical Engineering). Belo Horizonte - MG: UFMG, 2009. 206 p.

SHAYANI, R. A.; OLIVEIRA, M. A. G. de; CAMARGO, I. M. de T. Políticas públicas para a Energia: Desaf desafios para o próximo quadriênio, Comparação do custo entre energia solar fotovoltaica e fontes convencionais. V Congresso Brasileiro de Planejamento Energético, Brasília, p. 9-12, 2006.

SILVA, E. P. da. Study of the feasibility of using photovoltaic solar energy to charge batteries for various purposes. Postgraduate programme (Alternative forms of energy). Lavras - MG: UFLA, 2010. 178 p.

SOLAR, How to install solar energy. Available at: <http://www.portalsolar.com.br/como-instalar-energia-solar.html>. Accessed on 16 April 2016 a.

SOLAR, How to prepare your building for solar energy. Available at:<http://www.portalsolar.com.br/como-preparar-a-sua-construcao-para-energia- solar.html>. Accessed on 16 April 2016 b.

SOUZA, R. L. S. de; SILVA, F. R. C. da; SILVA, N. F. da. Harnessing solar energy in public lighting in Florianópolis. Ilha Digital Magazine, v. 2, p. 69-74, 2010.

TEIXEIRA, A. de A.; CARVALHO, M. C.; LEITE, L. H. de M. Feasibility analysis for the implementation of a residential solar energy system. e-xacta, v. 4, n. 3, 2012.

TEÓFILO, A. F. G.; SOUZA, M. de; MESQUITA, R. P. Electricity generation with photovoltaic solar cells for low-income rural populations. V Encontro de Energia no Meio Rural, p. 2-3, 2004.

VICHI, F. M.; MANSOR, M. T. C. Energy, environment and economy: Brazil in the global context. Chem. Nova, v. 32, n. 3, p. 757-767, 2009.

WANDERLEY, A. C. F. Perspectives for the insertion of photovoltaic solar energy in the generation of electricity in Rio Grande do Norte. Postgraduate Programme (Electrical and Computer Engineering). Natal - RN: UFRN, 2013. 126 p.

Buy your books fast and straightforward online - at one of world's fastest growing online book stores! Environmentally sound due to Print-on-Demand technologies.

Buy your books online at
www.morebooks.shop

Kaufen Sie Ihre Bücher schnell und unkompliziert online – auf einer der am schnellsten wachsenden Buchhandelsplattformen weltweit! Dank Print-On-Demand umwelt- und ressourcenschonend produzi ert.

Bücher schneller online kaufen
www.morebooks.shop

Printed by Books on Demand GmbH, Norderstedt / Germany